Explorers in North America

Solving Addition and Subtraction Problems Using Timelines

Kerri O'Donnell

PowerMath™

The Rosen Publishing Group's
PowerKids Press™
New York

Published in 2004 by The Rosen Publishing Group, Inc.
29 East 21st Street, New York, NY 10010

Book Design: Michael Tsanis

Photo Credits: Cover, p. 22 © Corbis; pp. 4, 12 (Jacques Cartier, Giovanni da Verrazano), 14, 18, 24, 26
(Meriwether Lewis, William Clark), 28 © Bettman/Corbis; p. 6 © Ted Spiegel/Corbis; p. 8 © Ewing
Galloway/Index Stock Imagery; p. 10 © Stock Montage/SuperStock; p. 16 © North Wind Picture Archives;
p. 20 © Museum of the City of New York/Corbis; p. 22 (inset) © Corbis; p. 30 © James L. Amos/Corbis.

Library of Congress Cataloging-in-Publication Data

O'Donnell, Kerri, 1972-
  Explorers in North America : solving addition and subtraction problems
using timelines / Kerri O'Donnell.
        v. cm. — (PowerMath)
Includes index.
Contents: Adventurous explorers — The Vikings arrive — The Europeans
set sail — Exploring Canada and New York Harbor — Exploring the mighty
Mississippi — Lewis and Clark — The move West.
  ISBN 0-8239-8987-9 (lib. bdg.)
  ISBN 0-8239-8898-8 (pbk.)
  6-pack ISBN: 0-8239-7426-X
  1. Addition--Juvenile literature. 2. Subtraction—Juvenile
literature. 3. North America—Discovery and exploration—Juvenile
literature. [1. Addition. 2. Subtraction. 3. North America—Discovery
and exploration.] I. Title. II. Series.
  QA115 .036 2004
  513.2'11—dc21
                                  2002156179

Manufactured in the United States of America

Contents

Adventurous Explorers

Throughout history, many people have traveled great distances to discover new lands. These explorers had different reasons for making these dangerous journeys. Explorers often set out with a particular goal—perhaps to extend their country's empire and increase its trade, or to gain wealth and fame for themselves. Some wanted to learn more about areas of the world people knew little or nothing about. Others wanted to experience the thrill of adventure. Some explorers traveled to distant lands to spread their religious beliefs.

This book will use timelines to show the history of exploration in North America. Timelines are tools that can help us organize important dates and events in the order that they happened.

Explorers called North America the "New World" because to them it was a strange land where they saw plants, animals, and people that looked much different from those in their homelands.

The Vikings Arrive

The Vikings were probably the first explorers to reach the New World. They came from what are now the northern European countries of Sweden, Denmark, and Norway.

The Vikings were **expert** sailors and fierce warriors. In the late 800s, they began to travel the North Atlantic Ocean in search of riches and new lands. Around 870 A.D., Vikings began to **migrate** to Iceland and settle there. In the 980s, Vikings in Iceland began to travel to Greenland to settle there. Around 986 A.D., a Viking ship making the trip to Greenland was blown off course, and the ship's captain saw a coastline to the west. This was probably the coast of North America, but the ship continued on to Greenland instead of exploring the new land.

The Vikings made their sea journeys in long, narrow ships made of wood. These ships were called "knarrs."

Leif Eriksson

870 A.D.	980s	986 A.D.	1000 A.D.	1014 A.D.
Vikings migrate to Iceland and establish settlements.	Vikings migrate from Iceland to Greenland.	Viking ship sights what is probably the eastern coast of North America.	Vikings led by Leif Eriksson reach North America and establish a colony.	Vikings abandon North American colony.

The Vikings in Greenland soon heard about the new land that had been discovered. By this time, most of Greenland's farmland had been claimed, so interest in the new land grew.

Around 1000 A.D., a Viking leader named Leif Eriksson and his men set out to find the new land. They landed along North America's eastern coast and formed a settlement they called "Vinland." Some **historians** believe that Vinland was located in what is now Newfoundland in Canada. Others think that it may have been in what is now Massachusetts or Maine.

More Vikings traveled to Vinland, and a colony was established. By 1014 A.D., however, the Vikings **abandoned** the colony, probably because of battles with the native people who lived in the area.

How many years after the Vikings first saw North America's coast did they establish a colony in North America? You can subtract 986 from 1000 to get your answer.

$$
\begin{array}{r}
0\ 9\ 9\ 10 \\
\cancel{1000} \\
-\ 986 \\
\hline
14 \\
\text{years}
\end{array}
$$

In 1000 A.D., 14 years after the Vikings first saw North America's coast, they established a colony there.

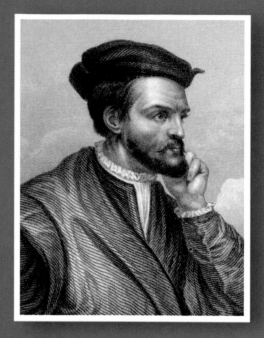

Giovanni da Verrazano   Jacques Cartier

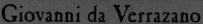

1492	1524	1534	1535
Columbus reaches North America.	Verrazano explores North America's eastern coast.	Cartier reaches North America.	Cartier explores St. Lawrence River.

Verrazano made two more voyages to the New World after his first trip in 1524. Some historians believe that he was killed by Native Americans in the Caribbean during the last of these voyages.

In the following years, it became clear that the large body of land that lay to the west across the Atlantic Ocean was not Asia, but a new land. During the 1500s, European explorers continued to travel to the New World, hoping to find a quick water route through the new land to Asia.

In 1524, the king of France sent an Italian explorer named Giovanni da Verrazano to North America to find a passage to Asia. Verrazano explored the eastern coast of North America from present-day North Carolina to Newfoundland in Canada, but did not find a passage.

A French explorer named Jacques Cartier reached North America in 1534, also looking for a passage to Asia. He did not find one, but he explored the St. Lawrence River in present-day Canada on his second voyage to the New World, in 1535.

Which explorer reached North America 42 years after Columbus did? You can add 42 to 1492 to get your answer.

$$1492 + 42 = 1534$$

Cartier reached North America in 1534—42 years after Columbus.

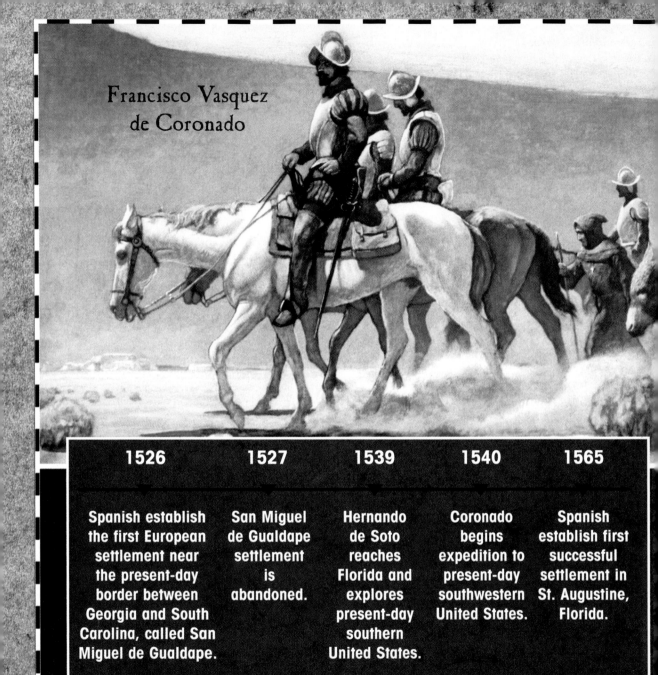

Francisco Vasquez
de Coronado

1526	1527	1539	1540	1565
Spanish establish the first European settlement near the present-day border between Georgia and South Carolina, called San Miguel de Gualdape.	San Miguel de Gualdape settlement is abandoned.	Hernando de Soto reaches Florida and explores present-day southern United States.	Coronado begins expedition to present-day southwestern United States.	Spanish establish first successful settlement in St. Augustine, Florida.

Around this same time, Spanish explorers were organizing expeditions that led them further **inland** into North America. In 1539, a Spanish explorer named Hernando de Soto and his crew of more than 600 men reached Florida's western coast, hoping to find gold. Their journey took them through what are today the southern U.S. states of Georgia, Alabama, Mississippi, and Arkansas. Although de Soto and his men did not find gold, they did become the first Europeans to see the Mississippi River.

In 1540, a Spanish explorer named Francisco Vasquez de Coronado set out on an **expedition** through the present-day states of Arizona, New Mexico, Texas, Oklahoma, and Kansas. His travels gave people back in Europe a better idea of how vast North America really was.

How many years passed between the time San Miguel de Gualdape was established and the time St. Augustine was settled? You can subtract 1526 from 1565 to get your answer.

$$
\begin{array}{r}
5\ 15 \\
1\cancel{565} \\
-\ 1526 \\
\hline
39 \\
\text{years}
\end{array}
$$

39 years passed between the time San Miguel de Guadalpe was established and the time St. Augustine was settled.

DES
SAVVAGES,
OV,
VOYAGE DE SAMVEL
CHAMPLAIN, DE BROVAGE,
fait en la France nouuelle,
l'an mil six cens trois:

CONTENANT

Les mœurs, façon de viure, mariages, guerres, & habitations des Sauuages de Canadas.

De la descouuerte de plus de quatre cens cinquante lieuës dans le païs des Sauuages. Quels peuples y habitent, des animaux qui s'y trouuent, des riuieres, lacs, isles & terres, & quels arbres & fruicts elles produisent.

De la coste d'Arcadie, des terres que l'on y a descouuertes, & de plusieurs mines qui y sont, selon le rapport des Sauuages.

A PARIS,
Chez CLAVDE DE MONSTR'ŒIL, tenant sa
boutique en la Cour du Palais, au nom de Iesus.

AVEC PRIVILEGE DV ROY

Exploring Canada and New York Harbor

In the early 1600s, European explorers were still searching the New World's waterways for a passage to the riches of Asia, but Europeans found that the New World had many of its own riches that could be traded for profit. They traded with the Native Americans, giving them European goods in exchange for valuable animal furs, which could be sold in Europe for very high prices.

In 1603, a French explorer named Samuel de Champlain reached present-day Canada to search for a route to Asia. He explored the northeastern coast of North America and founded the city of Quebec as a fur-trading post in 1608. In 1609, Champlain became the first European to reach a large lake in present-day New York and Vermont that was later named after him—Lake Champlain.

This title page is from a book Champlain wrote when he returned to France. In the book, Champlain wrote about his explorations and the Native Americans he encountered during his travels.

Henry Hudson

1603	1608	1609	1610
Champlain reaches present-day Canada.	Champlain establishes the city of Quebec.	Henry Hudson reaches present-day New York Harbor.	Hudson reaches Hudson Bay.

In 1609, an English explorer named Henry Hudson was hired by Dutch merchants to find a water route through the New World to Asia. He attempted to sail around the North Pole, but failed and went south instead. Hudson reached present-day New York Harbor, and then traveled north up a river that was later named after him—the Hudson River. He incorrectly believed that the river was a passage to the Pacific Ocean and Asia. Hudson claimed the land around the harbor and along the river for the Dutch.

The following year, Hudson set out on an expedition for the English to explore parts of what is now Canada. During his voyage, Hudson reached a large bay that he wrongly thought was the Pacific Ocean. This bay in northeastern Canada is now called Hudson Bay.

According to the timeline, what happened 5 years after Champlain reached present-day Canada? You can add 5 to 1603 to get your answer.

$$1603 + 5 = 1608$$

5 years after Champlain reached Canada, he established the city of Quebec.

An early map of New Amsterdam

Van Nieuw Engelandt. 21

t' Fort nieuw Amsterdam op de Manhatans

1613	1614	1624	1626
Adriaen Block reaches Manhattan Island.	Adriaen Block explores the Connecticut River.	Dutch send first permanent colonists to New York.	Dutch colonists build New Amsterdam on Manhattan Island.

A Dutch explorer and trader named Adriaen Block reached Manhattan Island in New York Harbor in the fall of 1613. Soon after reaching Manhattan, Block's ship was destroyed by fire. He and his crew built a small settlement on Manhattan Island to wait out the winter and lived there through the spring of 1614 while they built a new ship. They were the first Europeans to live on Manhattan Island.

After Block and his crew left New York Harbor that spring to return home, they sailed up the Connecticut River and claimed the land around it for the Dutch as part of the Dutch colony of New Netherland.

In 1624, a Dutch trading company sent the first permanent colonists to New York. Two years later, these colonists built New Amsterdam at the tip of Manhattan Island.

How many years passed between the time Adriaen Block reached Manhattan Island and the time New Amsterdam was built? You can subtract 1613 from 1626 to get your answer.

$$
\begin{array}{r}
1626 \\
- 1613 \\
\hline
13 \\
\end{array}
$$
years

13 years passed between the time Block reached Manhattan and the time New Amsterdam was built.

Mississippi River

A Member of the
Kickapoo Tribe

Exploring the Mighty Mississippi

The Mississippi River has played an important part in the history of North American exploration. It is one of North America's major rivers and is the second longest river in the United States. The Mississippi starts in Minnesota and flows about 2,340 miles (3,766 kilometers) south into the Gulf of Mexico, making it one of the largest inland waterways.

Hundreds of years ago, Native Americans living around the Great Lakes region told European explorers about a large waterway that flowed into a great sea. They called the river "Mississippi," which meant "big river." During the 1500s and 1600s, Spanish and French explorers sailed their boats up and down the Mississippi to explore the New World.

Native Americans in the upper Mississippi valley gave the Mississippi River its name. These Native Americans belonged to the Kickapoo, the Ojibway, the Illinois, and the Santee Dakota tribes.

Louis Jolliet and Jacques Marquette

1541	1673	1682
Hernando de Soto crosses Mississippi River near present-day Memphis, Tennessee.	Louis Jolliet and Jacques Marquette explore Mississippi River.	La Salle explores Mississippi River and claims surrounding land for France.

Hernando de Soto had been the first European to reach the Mississippi River in 1541. De Soto and the explorers that followed in his footsteps thought the river might flow into the Pacific Ocean.

In 1673, a French explorer and fur trader named Louis Jolliet set out to chart the Mississippi. He was accompanied by a priest named Jacques Marquette, who had learned the Native Americans' languages by living and working among them. The men set out from Lake Michigan, sailing through present-day Wisconsin to the Mississippi River. They soon found that the Mississippi flowed not west, but south into the Gulf of Mexico.

Later, around 1682, a French explorer named La Salle sailed the Mississippi, claiming the Mississippi valley for France.

How many years passed between the time de Soto crossed the Mississippi and the time Jolliet and Marquette explored the Mississippi? You can subtract 1541 from 1673 to get your answer.	$\begin{array}{r} 1673 \\ -\ 1541 \\ \hline 132 \\ \text{years} \end{array}$	132 years passed between the time de Soto first crossed the Mississippi and the time Jolliet and Marquette explored the river.

Meriwether Lewis

The Trail
of Lewis and Clark

William Clark

Lewis and Clark

European exploration led to the **colonization** of much of the New World. By the late 1700s, England had established 13 colonies along North America's eastern coast. English colonists soon decided they no longer wanted to be ruled by England and declared their independence in 1776. The colonies became the United States of America.

In 1803, the United States bought a large area of land called the Louisiana Territory from France. This was called the Louisiana Purchase. In 1804, two men named Meriwether Lewis and William Clark set out from Missouri on an expedition through this land to the Pacific Northwest. They wanted to find a route between the Atlantic Ocean and the Pacific Ocean, and to establish the boundaries of the territory and the Oregon region. They began by sailing up the longest river in the United States, the Missouri River, and stopped for the winter in present-day North Dakota.

Meriwether Lewis and William Clark knew each other for many years before they began their expedition. Lewis had served under Clark in the U.S. Army.

Sacagawea

1775	1776	1783	1804	1805
American Revolution begins.	English colonies in North America declare their freedom from England.	American Revolution ends.	Lewis and Clark begin their expedition to the Pacific Northwest.	Lewis and Clark reach the Pacific Ocean.

During the winter Lewis and Clark spent on the Great Plains, they met a Native American woman named Sacagawea, who became an **interpreter** for them. In the spring of 1805, they began traveling once again and eventually reached the Rocky Mountains in present-day Idaho. This part of the journey was very difficult—many of their horses lost their footing in the steep mountains and died, and food became **scarce** as the cold weather set in once again.

Lewis and Clark finally left the mountains behind them and made their way to the Columbia River, which they used to reach the Pacific Ocean in November 1805. For hundreds of years, explorers had been looking for a route to the Pacific. Lewis and Clark had finally found one, but it was much longer and more dangerous than people had expected.

How many years passed between the time the colonies declared their freedom from England and the time Lewis and Clark reached the Pacific Ocean? You can subtract 1776 from 1805 to get the answer.	7 9 15 1 8̶ 0̶ 5̶ − 1776 ——— 29 **years**	Lewis and Clark reached the Pacific Ocean 29 years after the colonies declared their freedom from England!

The Move West

The explorations of Lewis and Clark led to the **westward expansion** of the population of North America. In 1841, settlers from the eastern part of North America began traveling across the country to start new lives in the west. Many of them followed the Oregon Trail, the longest land trail to the Pacific Northwest. The Oregon Trail covered 2,000 miles (3,200 kilometers) as it made its way through North America's prairies, deserts, and mountains.

Much had happened between the time the Vikings had arrived in North America over 800 years before and the great migration west in the 1840s. The brave travels of all the adventurous people who explored the "New World" led to the settlement of an entire continent.

Glossary

abandon (uh-BAN-duhn) To give up something completely.

colonization (kah-luh-nuh-ZAY-shun) The act of establishing settlements in a distant land.

expedition (ek-spuh-DIH-shun) A journey made for a special purpose, often to learn about new places.

expert (EK-spurt) Someone who knows a lot about something.

historian (hih-STOR-ee-uhn) Someone who studies the past.

inland (IN-luhnd) Away from the coast.

interpreter (in-TUHR-pruh-tuhr) Someone who changes words in one language into words in another language.

migrate (MY-grayt) To move from one place to another place.

scarce (SKAIRSS) Hard to get. Rare.

westward expansion (WES-twuhrd ik-SPAN-shun) The movement of pioneer settlers to the western part of North America.

Index